Henrique Gabriel Rovigatti Chiavelli

Time and methods study in the foaming sector

Henrique Gabriel Rovigatti Chiavelli

Time and methods study in the foaming sector

At a Magnetic Mattress company

ScienciaScripts

Imprint
Any brand names and product names mentioned in this book are subject to trademark, brand or patent protection and are trademarks or registered trademarks of their respective holders. The use of brand names, product names, common names, trade names, product descriptions etc. even without a particular marking in this work is in no way to be construed to mean that such names may be regarded as unrestricted in respect of trademark and brand protection legislation and could thus be used by anyone.

Cover image: www.ingimage.com

This book is a translation from the original published under ISBN 978-3-330-76435-4.

Publisher:
Sciencia Scripts
is a trademark of
Dodo Books Indian Ocean Ltd. and OmniScriptum S.R.L publishing group

120 High Road, East Finchley, London, N2 9ED, United Kingdom
Str. Armeneasca 28/1, office 1, Chisinau MD-2012, Republic of Moldova, Europe
Managing Directors: Ieva Konstantinova, Victoria Ursu
info@omniscriptum.com

Printed at: see last page
ISBN: 978-620-8-50639-1

DEDICATORY

I dedicate this work to God, who gave me strength and enlightened me in the most difficult times, and to my parents for always encouraging me to study and never making me give up.

"A mind that is open to a new idea will never return to its original size."

Albert Einstein (1879 - 1955)

ACKNOWLEDGEMENTS

I would firstly like to thank God for always being present in my life, guiding me and directing my path every day so that I can achieve my goals in life and become a better person.

I would like to thank my parents who have always been present in my life, educating me, comforting me at difficult times in my life, being patient with me, and who have always encouraged me to dedicate myself to my studies.

I would like to thank my older brother Lucas, who was like a father to me in the city of Maringá because he always looked after me and kept me company, guided my path and was always present in my life, helping me with everything I needed.

I would like to thank my sister-in-law Cristina who, together with my brother, helped me a lot in Maringá when I needed it and with her group of friends made my life in Maringá more lively, making me forget the difficulties of everyday life.

I would like to thank my advisor Professor Carlos Antonio Pizo, who agreed to be my advisor and with all his professionalism and knowledge contributed greatly to this work, as well as dedicating himself to guiding me, informing me, correcting me and contributing in various ways to this work.

I would like to thank all my teachers, from my first years at school, who have contributed in some way to my education.

I would like to thank my friends, both the ones I have in the city of Marília and the ones I made during my degree, especially my group of friends who together managed to reach the end of another cycle of life and with whom university life was much better.

Finally, I would like to thank the company where this work was carried out, which allowed it to be done, all the people involved in this work, from the employees, who accepted the work carried out in their sector, to the production coordinators who guided me in the company, contributing to the realisation of this work.

SUMMARY

In an increasingly competitive market, companies are increasingly looking to reduce costs, improve product quality and increase productivity. With this in mind, they are looking for innovation, technology and continuous improvement in their production processes. This paper presents the study of times and methods as a tool for seeking improvements in a foaming sector in a magnetic mattress company, where 5 different densities are manufactured. With this study, a standard time was determined for each density and the sector's production capacity. By means of chronoanalysis, it was possible to verify some improvements, which were applied to the sector and determined a new standard time for each foam block density and a new production capacity, resulting in an increase from 25 foam blocks per day to 45, an increase of 80% and a reduction in daily overtime.

Keywords: Chronoanalysis, standard time, method and foaming.

SUMMARY

CHAPTER 1

INTRODUCTION

Industries have always been looking for new working methods to improve production processes and consequently increase production. One example is Henry Ford, who contributed his mass production model through the assembly line inspired by continuous flow production methods.

According to Barnes (2008), this quest to determine the fastest and most efficient method for carrying out an operation led Frederick W. Taylor, Frank Bunker Gilbreth and his wife Lillian Moller Gilbreth to develop what is now known as the study of times and methods at the beginning of the 20th century. The first, Frederick Taylor, introduced the study of time and the determination of standard times for processes. The Gilbreths are credited with developing the study of movements and improving working methods.

Although the study of Time and Methods was developed at the beginning of the 20th century, this process improvement technique only arrived in Brazil in the 1950s, brought by multinationals. It was implemented mainly in the ABC region of São Paulo in metalworking companies and car manufacturers. The clothing and footwear industries were the first to implement this study in their processes.

For Slack et al. (2009), method study is the systematic recording and critical examination of existing methods as a means of developing and applying easier and more convincing final methods of reducing costs.

The search for improvements, the advance of technology and fierce competition are increasingly constant in this globalised world where companies want to optimise their processes, whether with an emphasis on improving production time or making them cleaner, by monitoring and analysing unwanted results such as waste of raw materials, failure to meet delivery deadlines, production below target, products not meeting customer needs, among others. In this context, producing goods on a larger scale with the aim of reducing production costs and, in turn, reducing production process time, consequently increases the volume to be produced in order to guarantee delivery to customers, which is essential in terms of reliability and delivery time. As a result, the study of times and methods is increasingly being used by companies.

The aim of this work was to use the study of times and methods to improve the working method of the foaming sector of a magnetic mattress company in order to eliminate overtime and increase daily production, so that this sector can meet demand.

1.1. Justification

This work was carried out in the foaming sector of a magnetic mattress company located in the city of Maringá-

PR.

This topic was chosen because the author was doing an internship in this area and the Planning, Programming and Production Control (PPCP) sector needed to define its real daily production capacity for foam blocks, since five different densities are produced because the mattress assembly requires several laminated layers of foam, so that programming can be carried out based on the standard times for each density.

Another factor in carrying out this study was to increase the production of foam blocks and eliminate the daily overtime worked by operators to meet the demand required by the programme.

1.2. Definition and delimitation of the problem

The problem to be analysed is the daily overtime in the foaming sector, which is unable to meet daily demand and consequently has a direct impact on production.

1.3. Objectives

1.3.1. General objective

The aim of this work is to carry out a time and methods study to determine production capacity and increase production in the foaming sector.

1.3.2. Specific objectives

The specific objectives of this work are:

- Analysing production methods;
- Determine the standard time for each foam block density;
- Calculate production capacity;
- Eliminate overtime;
- Increase production.

CHAPTER 2

LITERATURE REVIEW

This chapter covers topics that contribute to the development of the work. The concepts of the study of times and methods, the definitions of timed times, synthetic times, work sampling and the importance of ergonomics and the work method for the study of times and methods will be presented.

2.1. Time and Methods Study

The study of time and motion has become one of the most effective tools in the field of engineering when it comes to determining work efficiency by setting parameters for production programmes and reducing industrial costs. Among these industrial costs are labour costs, which need to be addressed in order to reduce the company's costs.

According to Peinado and Graeml (2007), the study of time is the determination of the time required to carry out a task, using a stopwatch, with the aim of determining the best and most efficient way to carry out a specific task and also to find a reference standard that will serve to:

- Determine the company's production capacity;
- Drawing up production programmes;
- Determine the cost of direct labour when calculating the cost of goods sold (COGS);
- Estimating the cost of a new product during its design and creation;
- Balancing production and assembly lines.

According to Barnes (2008), the study of time began in 1881 at the Midvale Steel Company with the work carried out by Frederick Taylor. He carried out studies to optimise the movement of materials, looking for methods to improve the techniques used by the company in order to increase production per worker. As a result, labourers began to perform their tasks more economically, eliminating unnecessary time. The study of movements originated in 1885 and was introduced by the couple Frank B. Gilbreth and his wife Lillian M. Gilbreth. He was an engineer and she was a psychologist who complemented each other in terms of knowledge, which led them to carry out work that involved understanding the human factor, as well as knowledge of materials, tools and equipment. Barnes (2008) explains that for several years the main objective of the field of movement and time studies was to establish standard times for use in wage incentive plans. Despite this, it has been realised that the study of movements is also a powerful tool for reducing costs, as it reduces the time it takes to carry out a given activity.

According to Barreto (1997), every operation in a factory demands time, from the most insignificant to the most complex of operations, so the aim of studying time and method is to provide company management with real knowledge of the time needed to produce something, thus having greater control over production processes.

According to Chiavenato (2000), for Taylor and his followers, the basic instrument for rationalising the work of workers was the study of time and movements. They found that work could be done better and more economically by analysing the work, i.e. by dividing and subdividing all the movements needed to carry out each operation performed by the workers. Taylor observed the possibility of breaking down each task operation into an ordered series of simple movements. Useless movements were eliminated, while useful movements were simplified, rationalised or merged with other movements to save the worker time and effort.

For Peinado and Graeml (2007), the study of times, movements and methods is linked to three important definitions in the business vocabulary: methods engineering, work design and ergonomics.

According to Toledo Jr (2004), the study of times and methods is the analysis of the material methods, tools and installations used and that will be used to carry out a job. The purpose of this analysis is to

- Finding the most economical way to carry out a job;
- Standardising methods, materials, tools and installations;
- Determine exactly how long it takes for a skilled person to carry out the work at a normal pace;
- To help workers learn a new method of working.

According to Toledo Jr (2004), these four topics, although distinct, cannot be separated, because if this were to happen, it could cause great damage to the study.

According to Martins and Laugeni (2005), three different methods can be used to carry out the Time and Methods study: the Timed Times method (Chronoanalysis); the Predetermined Times method (Synthetic Times); and the Work Sampling method. The choice of the ideal method will depend on where it will be applied.

2.1.1. Timed Times (Chronoanalysis)

Martins and Laugeni (2005) state that:

> Timekeeping is one of the most common methods used in industry to measure work. Despite the fact that the world has undergone considerable changes since the time when F.W. Taylor structured Scientific Management and the study of timed work in order to measure individual efficiency, this methodology is still widely used to establish standards for production and

industrial costs.

According to Martins and Laugeni (2005), production times will vary little when the production line is automated, but the more human intervention there is in production, the greater the variation in times. It will therefore be more difficult to measure times correctly, as each operator has different skills, strengths and wills. Standard production times are important for:

- Establishing parameters for production programmes for factory planning, making effective use of available resources and also for evaluating production performance in relation to the existing standard;
- To provide data for determining standard costs, for surveying manufacturing costs, determining budgets and estimating the cost of a new product;
- Provide data to study production structures, compare production routings and analyse capacity planning.

In addition to these purposes, Barnes (2008) also mentions that standard time can be useful to use as a basis for incentive payments for direct labour, to determine the efficiency of machines, the number of machines a person can operate, the number of men needed to run a group, and as an aid to balancing assembly lines and conveyor-controlled work.

According to Krick (1976 **apud** Camarotto, 2005), the standard time is the time needed to complete a cycle of an operation when carried out using a given method, at a certain arbitrary working speed, with the expectation of delays and delays independent of the value.

Camarotto (2005) mentions that standard time is the time used to determine work capacity in production centres where operators are involved, either in exclusively manual activities or in man-machine interaction.

According to Ritzman and Krajewski (2004 **apud** Peinado and Graeml, 2007), production capacity is the highest level of output that a company can reasonably maintain using realistic employee working hours and the equipment currently installed.

According to Slack et al (2009), production capacity is defined as the maximum level of value-added activity in a given period of time that the process can carry out under normal operating conditions.

For Gaither & Frasier (2001 **apud** Peinado and Graeml, 2007), capacity is the highest level of production that a company can maintain within the framework of a realistic work schedule, taking into account a period of normal downtime and assuming sufficient availability of inputs to operate existing machinery and equipment.

According to Peinado and Graeml (2007), the equipment used to carry out the time study follows:

- **Centesimal hour chronometer:** this is the most commonly used. One revolution of the larger hand corresponds to 1/100th of an hour, or 36 seconds. It can be found in specialised shops. Timing can also be done with an ordinary stopwatch (sexagesimal time), but the time measured must be converted to centesimal time, as shown in Table 1.

Time measured with an ordinary stopwatch	Time converted to the centesimal system	Calculation
1 minute and 10 seconds	1.17 minutes	1 + 10/60= 1,17
1 minute and 20 seconds	1.33 minutes	1 + 20/60 = 1,33
1 minute 30 seconds	1.50 minutes	1 + 30/60 = 1,50
1 hour, 47min and 15sec.	1.83 hours	1 + 47/60 + 15/3600 = 1,79

Table 1: Conversion from sexagesimal to centesimal time Source: Peinado and Graeml (2007)

- Camcorder: A camcorder is used to measure the time needed to carry out the task. The use of a video camera has the advantage of faithfully recording all the movements made by the operator and, if used correctly, can eliminate the psychological tension that the operator feels when being observed directly by a chrono-analyst;
- **Clipboard: This is** used to support the stopwatch and the observation sheet, so that the chrono-analyst can write down his time-taking while standing up;
- **Observation sheet:** This is a document in which the times and other observations relating to the timed operation are recorded.

According to Martins and Laugeni (2005), the first step is to talk to and inform everyone involved about the type of work that will be carried out, seeking the co-operation of those in charge and the operators in the sector. Next, the method of the operation must be defined and divided into time elements, i.e. the process to be timed will be analysed and, if necessary, improved, thus defining the best method for carrying out the process. Once the process has been standardised, it will be divided into the activities that the workers carry out in order to time each activity.

Barnes (2008) also mentions that operators should be informed in advance of the time study being carried out, as well as being made aware of the study's objectives.

Peinado and Graeml (2007) also point out that the total operation whose standard time is to be determined must be divided into parts so that the working method can be measured accurately. Care must be taken not to split the operation into too many or too few elements. Some general rules for this breakdown are:

1. Separate the work into parts, so that they are as short as possible, but long enough to be measured with a stopwatch;
2. Operator actions, when independent of machine actions, must be measured separately;

3. Define delay caused by the operator and the equipment separately.

Martins and Laugeni (2005) state that the main purpose of the division of elements is to verify the working method.

According to Barnes (2008), in order to carry out a time study, eight steps must be followed:

1. Obtain and record information about the operation and the operator under study;
2. Divide the operation into elements and record a full description of the method;
3. Observe and record the time spent by the operator;
4. Determine the number of cycles to be timed;
5. Assess the operator's rhythm;
6. Check that a sufficient number of cycles have been timed;
7. Determine tolerances;
8. Determine the standard time for the operation.

According to Barnes (2008) data collection and recording has three common methods for reading the stopwatch: (1) continuous reading; (2) repetitive reading and (3) accumulated reading. The first two methods are most commonly used.

1. **Continuous Reading:** In this method, timing starts at the beginning of the first element and keeps the stopwatch running until the last element, but at the end of each element the time must be recorded on the observation sheet in front of each respective element, so the time of each element, starting from the second element, will be given by subtracting the next element from the previous element;
2. **Repetitive reading:** In this reading method, the timing starts with the first element and at the end of each element the stopwatch must return to zero, starting each element with the stopwatch set to zero, noting the value on the check sheet before resetting the stopwatch;
3. **Accumulated Reading:** This method allows direct reading of the time for each element using two stopwatches. These chronometers are mounted together on the board and linked by a lever mechanism, so that when the first chronometer is started, the second automatically stops and vice versa. This method has become obsolete with the advent of digital stopwatches, which make timekeeping easier and the assembly of the mechanism unfeasible.

According to Martins and Laugeni (2005), in order to determine the standard time for a part or operation, 10 to 20 timings should be carried out. However, there is a mathematical way to determine the number of timings to be carried out, which is through Equation 1.

$$n = \left(\frac{z * R}{E_r * d_2 * \bar{X}}\right)^2 \qquad (1)$$

Where:

= = number of cycles to be timed;

z = coefficient of the standard normal distribution for a given probability;

= = sample amplitude;

E_r = relative error;

d_2 = coefficient based on the number of preliminary timings; X = sample average.

According to Martins and Laugeni (2005), in order to use Equation 1, the operation should be timed 5 to 7 times beforehand and the results obtained should be the mean X and the amplitude /? The probability values should be set between 90% and 95%, and the relative error should vary between 5% and 10%.

The typical values of the z and d_2 coefficients used in the calculations are shown in Table 1 and Table 2, respectively.

Table 1: Normal distribution z coefficient

Probability	90%	91%	92%	93%	94%	95%
z	1,65	1,70	1,75	1,81	1,88	1,96

Source: Martins and Laugeni (2005)

Table 2: Coefficient d_2 of normal distribution

No. of cycles (n)	2	3	4	5	6	7	8	9	10
d_2	1,128	1,693	2,059	2,326	2,534	2,704	2,847	2,970	3,078

Source: Martins and Laugeni (2005)

According to Barnes (2008), after determining the number of cycles to be timed, the operator's pace during the study must be assessed, known as the Pace Factor (PF), which is the most important and difficult phase of the time study, in which the time study analyst compares the pace of the operator under observation with their own concept of normal pace. Subsequently, this pace factor will be applied to the Timed Time (TC) in order to obtain the Normal Time (TN) for this task, which is given by Equation 2.

$$TN = TC \ x \ FR \qquad (2)$$

Where:

TN = Normal Time;

TC = Timed Time;

FR = Rhythm Factor.

According to Slack et al. (2009), evaluating the pace of observed times is the process of assessing the worker's work speed in relation to the observer's concept of the speed corresponding to standard performance. The observer can take into account, separately or in combination, one or more factors necessary to carry out the work, such as speed of movement, effort, dexterity, consistency, among others.

For Peinado and Graeml (2007) the evaluation of the operator's speed is the process by which the chrono-analyst compares the rhythm of the operator under observation with their own concept of normal rhythm. The operator's normal operating speed is assigned a speed rate, or rhythm, of 100 per cent. Above-normal speeds have values greater than 100 per cent and below-normal speeds have values less than 100 per cent.

Barnes (2008) highlights six methods for evaluating the operator's pace, but this paper will focus on three methods:

1. **Evaluation of pace through skill and effort:** this system is based on time studies, with their standards expressed in points or "B". The assessment is made

   taking into account the operator's skill (the degree of ease with which the operator carries out an activity) and effort (the operator's commitment to carrying out the activity as quickly and as well as possible), using a standard table (Table 3) of tolerances for fatigue (60 points corresponds to standard execution). In this system, a job carried out at a normal pace produces 60 B per hour and the average performance under an incentive system ranges from 70 to 80 B per hour;
2. **Rhythm performance:** this is the most widely used system in the US, evaluating a single factor between operator speed, rhythm or time. This system uses previous time records to establish normal standards. For example, if a speed of 5 kilometres per hour is considered normal (100%), 6 kilometres per hour represents 120% in the pace assessment. The estimate can be made for one element or for a complete cycle of elements;
3. **Westinghouse system for assessing pace:** this system assesses four factors: (1) skill; (2) effort; (3) conditions and (4) consistency, where there is a table that will provide numerical values for each factor as shown in Table 3. In this system you add up the value of each factor, add one more and multiply by the timed time, as in the example below:

   Example: Timed time = 12mim;

   Skill = Excellent B2 = +0.08;

   Effort = Good C1 = +0.05;

   Conditions = Ideal = 0.06;

   Consistency = Perfect = +0.04

   Total = +0.08 + 0.05 + 0.06 + 0.04 = 0.23 + 1 = 1.23

Normal Time (NT) = 12 x 1.23 = 14.76min.

Table 3: Rhythm evaluation system

SKILLS			EFFORT		
+0,15	A1	Superhabit	+0,13	AI	Excessive
+0,13	A2		+0,12	A2	
+0,11	BI	Excellent	+0,10	BI	Excellent
+0,08	B2		+0,08	82	
+0,06	CI	Good	+0,05	Cl	Good
+0,03	C2		+0,02	C2	
0,00	0	Medium	0,00	□	Medium
-0,05	EI	Regular	-0,04	EI	Regular
-0,10	E2		-0,08	E2	
-0,16	F1	Weak	-0,12	F1	Weak
-0,22	F2		-0,17	F2	
CONDITIONS			**CONSISTENCY**		
+0,06	A	Ideal	+0,04	A	Perfect
+0,04	B	Excellent	+0,03	B	Excellent
+0,02	C	Good	+0,01	C	Good
0,00	D	Average	0,00	D	Average
-0,03	E	Regular	-0,02	E	Regular
-0,07	F	Weak	-0,04	F	Weak

Source: Barnes (2008)

According to Barnes (2008), it is important to check that the number of timed cycles has been sufficient in order to determine tolerances, because normal time has no tolerances and it is not expected that a person will work all day without any interruptions, so the operator may spend part of their time on personal needs to rest or for reasons beyond their control. Tolerances for production interruptions can be categorised as follows:

1. **Personal tolerance:** According to Peinado and Graeml (2007), this refers to the body's physiological needs, and one way of determining the length of this tolerance would be to use work sampling or continuous monitoring. However, for light work with eight-hour days, without a rest break except for lunch, the average downtime varies from 10 to 24 minutes, i.e. from 2% to 5% of the working day, remembering that heavier work and hot and humid environments require more time for personal needs;
2. **Tolerance for fatigue:** According to Martins and Laugeni (2005), tolerances for fatigue range from 10 per cent (light work in a good environment) to 50 per cent of the time (heavy work in unsuitable conditions).

   a tolerance ranging from 15 to 20 per cent of the time for normal work carried out in a normal environment, for industrial companies;
3. **Tolerance for waiting:** According to Barnes (2008), waiting during production can be avoidable or not, and waits made intentionally by the operator should not be taken into account when determining the standard time. Waiting can occur due to stoppages such as: machine maintenance, tool breakage, interruptions from supervisors, tool replacement, etc. This tolerance

can be determined through continuous studies or work sampling.

According to Camarotto (2005), in addition to these three tolerances there is also the special tolerance, which is given due to special situations such as: lack of adequate training for the job, labour turnover, adverse social conditions, so estimating this tolerance is difficult to measure, and it is necessary to use work sampling to determine it.

According to Martins and Laugeni (2005), in addition to these methodologies for determining tolerances, they can also be calculated according to the time allowances that the company is willing to grant. In this case, the percentage of time p granted is determined in relation to the daily working time and the tolerance factor is calculated using Equation 3.

$$FT = \frac{1}{(1-p)} \tag{3}$$

Where:

FT = Tolerance Factor;

p = given break time divided by working time.

There is no satisfactory way of measuring tolerance, as this will depend on the nature and environmental conditions of the work. Figure 1 shows the tolerances proposed by Niebel (1992, apud Peinado and Graeml, 2007).

DESCRIPTION	%	DESCRIPTION	%
A. Unvarying tolerances:		4. Poor lighting:	
]. Tolerances for personal needs	5	at well below the recommended level	0
2. Basic fatigue tolerances	4	b. well below what is recommended	2
B. Variable tolerances:		c. multo Inadeguada	5
1. tolerance for standing	2	5. Atmospheric conditions	0-10
2. Tolerance of posture		(heat and umldadel - variâve !s	
a. largely uninhabited	0	6. Careful attention	
b. uninhabited (recurved}	2	reasonably fine work	0
c. multo desabitado (lying down, stretched out)	1	b. fine or precision work	2
3. Use of force or muscular energy		c. fine or very precise work	5
íerguer. pull or lift)		7. Noise level:	
Weight lifted in gullos		to be continued	0
2,5	0	b. intermittent - high volume	2
5,0	2	c. Intermittent - very high volume	5
7,5	2	d. high timbre-high volume	5
10,0	3	&. Mental stress	
12,5	4	a. reasonably complex process	1
15.0	5	b. complex, comprehensive process	4
17.5	7	c. very complex process	8
20,0	0	9. Monotony:	
22.5	11	i low	0
25.0	13	b. average	1
27.5	17	c. high.	4
30.0	22	10. Degree of boredom	
		a. rather boorish	0
		b. tedious	2

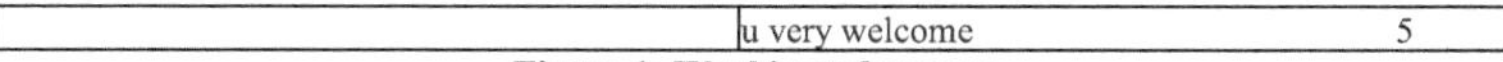

Figure 1: Working tolerances
Source: Peinado and Graeml (2007)

According to Barnes (2008), after determining the Tolerance Factor (FT), the Standard Time (ST) must be calculated, which according to Peinado and Graeml (2007), is calculated by multiplying the normal time by the tolerance factor, as shown in Equation 4.

$$TP = TN\ x\ FT \quad (4)$$

Where:

TP = Standard Time;

TN = Normal Time;

FT = Tolerance Factor.

2.1.2. Predetermined Times (Synthetic Times)

According to Peinado and Graeml (2007), by carrying out a time study, the company will have stored the elementary times that are common to various functions. The synthetic time methodology has an advantage over chrono-analysis, because it is possible to determine the

the execution time of an activity or a product without the need for timing, i.e. you can determine the standard time for a given product without it having been launched or for a job without it having started.

According to Barnes (2008), the main uses of synthetic time systems are divided into two classes, as shown in Table 2.

Method evaluation	Establishing standard times
Improving existing methods.	Direct use of synthetic times to establish standard times.
Evaluation of proposed methods before production begins.	Compilation of standard data and formulas for specific classes of work to speed up the establishment of standard times.
Evaluation of tool, device and equipment projects.	Verification of standards established by time studies.
Product design assistance.	Standard time auditing.
Training of supervisory staff to guide them through the study of movements and times.	

Table 2: Main uses of synthetic time systems Source: Barnes (2008)

According to Barnes (2008), there are nine synthetic time systems. Due to the lack of published information and the specificities of each method made or adapted for each particular company, it is impossible to know how many different synthetic time systems may be in use in organisations. However, despite the high degree of specification, all the methods have a lot in common. Of the nine systems, four synthetic time systems are described: Synthetic times for assembly operations; the Basic Time Measurement (BTM) system; the Labour Factor system and the Methods Time Measurement (MTM) system, the last two being the main synthetic time systems, but MTM is the most widely used.

2.1.2.1. Synthetic times for assembly operations

According to Barnes (2008), this system was developed by Harold Engstrom and his colleagues at the General Electric Company plant in Bridgeport. This system was used to estimate labour costs for new products and establish standard times. It was a system designed for assembly operations in the household appliance division, not intended for universal application.

2.1.2.2. Basic Time Measurement (BTM) system

According to Barnes (2008), this is called the basic motion time study and was developed by Ralph Presgrave, G.B. Bailey and other members of J.D. Woods and Gordon, LTD, in Canada. It was first used in 1950. This system takes into account the distance travelled, visual attention, the degree of precision required in grasping or positioning, the force required in handling the weight and the simultaneous execution of two movements.

2.1.2.3. Labour Factor System

According to Barnes (2008), this system was one of the first synthetic time methods to be generalised and was first applied in 1938. Using this method, it is possible to determine the normal time or selected time for manual tasks using synthetic data. In this system, a basic movement is defined as one that involves the minimum difficulty or precision for a given distance and combination of limbs. The labour factor is a unit used as an index of the additional time required over and above the basic time when movements are executed involving the following variables: (a) manual control (involving four main factors that affect the execution time of manual movements, such as the body member used, the distance travelled, the manual control required and the weight or resistance encountered) and (b) weight or resistance (where the effect of weight varies with the body member used and the sex of the operator.

2.1.2.4. Methods Time Measurement (MTM) system

According to Barnes (2008), this system was developed from the industrial operations film studio, and its standard times were published in 1948. This method is based on analysing any manual operation or method into the basic movements required for its execution, associating each movement with a synthetic time determined by the nature of the movement and the conditions under which it is executed. The unit of time used in this system is the hundred thousandth of an hour (0.00001 h), with a Time Measurement Unit (TMU) worth 0.0006 minutes.

According to Barnes (2008), there are nine fundamental movements which are divided into: reaching, moving, turning, grasping, positioning, releasing, dismounting, eye movements and body, leg and foot movements.

Table 3 explains each fundamental movement.

Fundamental Movements	Explanation and detailing of each fundamental movement
Reach	It is the basic element used when the main purpose is to transport the hand or finger to a destination. 0 time to reach varies with the condition, distance travelled and type of reach. This movement is categorised into five classes: 1. **Case A:** for an object in a defined location, for an object in the other hand or on which the other hand rests; 2. **Case B:** for an object whose general location is known. This location may vary slightly from cycle to cycle; 3. **Case C:** for objects located in a group of objects; 4. **Case D:** for a very small object or when precision gripping is required; 5. **Case E:** to an undefined location in order to balance the body, for the next movement or clearing the way;
Move	It is the basic element used when the purpose is to transport the object to a destination. 0 movement time is affected by the condition, distance travelled in movement and weight factor. There are three classes: 1. **Case A:** object to the other hand or from a meeting to a stop; 2. **Case B:** object for approximate or undefined localisation; 3. **Case C:** object for exact localisation.
Rotate	This is the rotational movement of the hand, wrist and forearm, with the forearm as its own axis. The time taken to rotate the hand can vary depending on the degree of rotation and the weight factor.
Grab	It is the basic element used to align, orientate and assemble an object with another object, where the movements used are such that classification in other basic elements is not justified. 0 positioning time is affected by the class of the fit, symmetry and ease of handling.
Position	It is the basic element for aligning, orientating and assembling an object with another object. 0 positioning time is affected by the class of the fit, symmetry and ease of handling.
Release	This is the basic element that refers to relinquishing the control exercised by the fingers or hand over the object. The two classifications of release are: (1) normal release, simple opening of the fingers; (2) contact release, in which the release begins and ends the instant the next reach begins.
Release	It is the basic element that refers to relinquishing the control exercised by the fingers or hand over the object. The two classifications of release are: (1) normal release, simple opening of the fingers; (2) contact release, in which the release begins and ends the instant the next reach begins.
Dismantle	This is the basic element that refers to relinquishing the control exercised by the fingers or hand over the object. The two classifications of release are: (1) normal release, simple opening of the fingers; (2) contact release, in which the release begins and ends the instant the next reach begins.
Time for the eyes	In most jobs, the time taken for the eyes to move and focus is not a limiting factor and consequently does not affect the operating time. However, when the eyes direct the movements of the hands or body, it is necessary to take eye movement times into account. There are two types of eye movement: (1) focussing time, which is the time taken to focus on a certain object and distinguish certain characteristics; (2) gaze movement time, which is affected by the distance between the points from which and at which the eyes move and by the distance measured at the perpendicular taken from the eye to the line of movement.
Movement of the body, leg and foot	The times of body, leg and foot movements are also classified in relation to the type of movement and the distance travelled. The types of movement can be foot movement, leg movement, stepping to the side, bending, kneeling, sitting, getting up from a sitting position and walking.

Table 3: Fundamental movements of the MTM system Source: Barnes (2008)

2.1.3. Labour Sampling

According to Barnes (2008, p. 416), "work sampling in its simplest form consists of making observations at occasional intervals of one or more operators or machines and recording when they are inactive or working".

According to Peinado and Graeml (2007), the purpose of work sampling is to obtain, with a statistically determined degree of reliability, an estimate of the duration of the activity with an admissible, statistically proven error.

According to Barnes (2008), the labour sampling methodology was first used in the British textile industry by L. H. C. Tippet. In the US, it was called the "wait ratio" and was applied in 1940. This method makes it possible to collect data in less time and at lower cost than other labour measurement methods.

Martins and Laugeni (2005) state that work sampling takes longer to collect data than the other methods. It involves estimating the proportion of time spent on a given type of activity, over a certain period, by means of instantaneous and randomly spaced observations.

According to Peinado and Graeml (2007), the idea of work sampling is to take a certain measurement on a small part of a population to be analysed and use this information to make inferences about the whole population. This methodology makes it possible to estimate the percentage of time a worker or machine spends on each activity. The method does not require continuous observation or timing of the activity.

According to Camarotto (2005), work sampling is based on the statistical method of sampling, the use of a random sample of a population, making it possible to know the percentage of time dedicated to work and rest in situations where the work cycle is very long or undergoes many variations in methods.

According to Barnes (2008), this method has three main uses:

1. **Waiting ratio:** measuring activities and waits of men and machines, for example, determining the percentage of a day in which a man works and does not work;
2. **Performance sampling:** to measure the working time and rest time of a person performing a manual task and to establish an index or level of performance for the same person during their working time;
3. **Labour measurement:** establishing a standard time for an operation, under certain circumstances.

Peinado and Graeml (2007) cite five applications for the use of this method:

- Determine the tolerance factor for the waiting time that can be incorporated into the standard time;
- Determine the degree of utilisation of machinery, apparatus and transport equipment and a worker's inactivity rates;
- Determine the indirect labour activity for cost apportionment (including the ABC system);

- Estimate the time spent on various activities;
- Estimating the standard time of an operation under certain circumstances.

In addition to these applications, Peinado and Graeml (2007) also mention that this methodology is widely used to determine the time spent on non-repetitive activities, which are more difficult to control and generally cover a wider range of activities, such as a quality inspector who can spend part of his time on important activities, such as contacting and determining specifications with suppliers, and part of his time on routine activities, such as checking part dimensions.

Barnes (2008) exemplifies this methodology by making a note of an operator when he is working and when he is inactive, so the percentage of the day that the operator was inactive is estimated by the ratio between the number of inactivity records and the total number of observations made. The relative error is a function of the number of observations made.

Peinado and Graeml (2007) suggest some rules for determining the proportion of time a person spends on their activities:

- Observations must be instantaneous, i.e. what matters is the activity that was observed at the moment the analyst "looked" at it, regardless of its duration;
- Observations should be made at random time intervals. No interval pattern should be followed;
- The number of observations must be sufficient to represent the population, according to the degree of reliability and error required by the study.

According to Peinado and Graeml (2007), Equation 5 is used to determine whether the number of observations is sufficient to represent the truth.

$$N = \left(\frac{z}{E_r}\right)^2 x \left(\frac{1-p}{p}\right) \quad (5)$$

Where:

N = number of observations required;

Z = normal distribution coefficient (values in Table 1);

Fr = acceptable relative error;

p= proportion of the activity studied in the set of all activities.

According to Martins and Laugeni (2005), work sampling has some advantages and disadvantages over timed times, as shown in Table 4.

Advantages	**Disadvantages**

Operations that are expensive to measure on a chronometer;	Not good for repetitive, short-cycle operations;
Simultaneous team studies;	It can't be as detailed as a stopwatch study;
The cost of a timekeeper is high;	The job configuration can change over time;
Long observations reduce the influence of occasional variations;	Sometimes you forget to record the working method;
0 operator does not feel closely observed.	The administration doesn't understand so well.

Table 4: Advantages and disadvantages of work sampling compared to timed times Source: Martins and Laugeni (2005)

2.2. Ergonomics and working methods

According to Barnes (2008), the main objective of work method design is to find the most efficient combination of men, machines, equipment and materials in the workplace. To do this, it is very important to determine which functions are carried out by man and which by machine. As ergonomics (a word derived from the Greek "ergon" meaning "work" and "nomos" meaning "laws or norms") aims to adapt tasks and the work environment to people's sensory, perceptual, mental and physical characteristics, the study of times and methods ends up resulting in improvements for the operator, increasing their productivity.

Camarotto (2005) explains that work study is the use of techniques, methods and work measurement in order to examine human work in all its aspects and thus investigate the factors that influence the efficiency and performance of the situation studied with the aim of improving comfort and safety in the execution of work and also increasing the productivity of processes. In order for a work study to be complete and effective, a work project is needed that will dimension the material and organisational resources needed to carry out a given set of tasks in a production centre.

According to Barreto (1997), before measuring a production time and considering it as a standard, the correct method for carrying out the work to be measured must be found.

According to Taylor (1990), there are five stages to finding the best method of carrying out a task:

- Select 10 to 15 operators who are skilled in the activity to be analysed;
- To study the exact cycle of elementary operations or movements that each of these men uses when carrying out the work under investigation, as well as the tools used;
- Using a stopwatch, study the time required for each of these basic movements and then choose the quickest ways to carry out the work phases;
- Eliminate all faulty, slow and useless movements;
- Bringing together in one cycle the best and fastest movements, as well as the best instruments.

According to Barnes (2008), when a product or service is being designed or developed there is a great opportunity to use process design and propose the best production methods and systems, but the "perfect method" does not exist, but there are always opportunities for improvement. Factors such as the volume and quality of the product, the type and price of raw materials and the availability of machinery and equipment may differ from those that were in force when production began. Therefore, there is always the opportunity to improve processes and methods, even to the point of redesigning the product itself and its components.

Barnes (2008) emphasises four possible solutions for selecting the preferred method for carrying out an activity:

- Eliminate all unnecessary work;
- Combining operations or elements;
- Change the sequence of operations;
- Simplify essential operations.

According to Másculo (2011), physical layout deals with the entire physical location of transformation resources, i.e. it determines where machines, equipment and personnel will be allocated, thus determining the shape and appearance of workplaces and how processes will flow. Therefore, physical layout interacts with Ergonomics directly in the aspects that concern the worker's activities, which is why a good physical layout provides:

- Security;
- Minimising distances;
- Good signposting;
- Comfort for operators;
- Ease of coordination.

According to Gomes and Másculo (2011), a work organisation model is conditioned by the qualifications of the workforce, the culture of the country and, above all, the capital labour relations in force, so the position of employees or collaborators varies from country to country.

According to Vidal (2011), Ergonomics can adapt and design the organisation so that it meets the challenges of the present (efficiency) and the future (sustainability), through the concepts of task, activity, variability and regulation. Ergonomics can therefore be understood as the study and implementation of new ways of solving problems posed by the way work is organised.

According to Iida (2005), ergonomics is the study of adapting work to humans, with the aim of reducing fatigue, stress, errors and accidents, providing safety, satisfaction and health for employees during their relationship with the production system.

According to Vidal (2011), the anthropometric study covers the methods and techniques that enable us to obtain a satisfactory set of measurements and conformations of the body or parts of the human body, so this is a necessary stage in defining the design of a workstation, being directly responsible for an important part of the dynamics of the movements for carrying out a particular activity. Therefore, if a workstation is poorly designed, the operator will certainly be forced to adopt forced postures at a given moment or perform a sequence of unbalanced movements in a dynamic configuration.

According to Vidal (2011), Ergonomics has been applied in the organisational field in:

- Process modelling to draw up scenarios and roadmaps for organisational change;
- Analysis of the requirements of the new organisational proposals in terms of capacities, limitations and other characteristics, specifying training needs and new competences;
- Building implementation roadmaps to avoid decapitalising or underutilising existing **know-how**, especially at the operational level;
- Expertise and accident prevention.

According to Iida (2005), there are two types of approaches to analysing the workplace: Taylorist and ergonomic. The Taylorist approach is based on the principles of the economy of movements, which is called the study of times and movements. This is based on a series of empirical findings accumulated since Taylor's time (1856-1915). Therefore, the best method is chosen based on the criterion of the least amount of time spent. To choose the best method, three steps should be followed: (1) develop the preferred method; (2) prepare the standard method and (3) determine the standard time. The ergonomic approach tends to develop workstations that reduce biomechanical and cognitive demands. This approach seeks to ensure worker satisfaction and safety and system productivity, as well as eliminating highly repetitive tasks.

CHAPTER 3

METHODOLOGY

Research is necessary for everyone, and according to Gil (2007, p. 42) "research has a pragmatic character, it is a formal and systematic process of developing the scientific method. The fundamental objective of research is to discover answers to problems through the use of scientific procedures".

To carry out this work, we first carried out a bibliographical survey of the theoretical foundations, using books, articles, case studies, academic papers and other sources.

According to Silva and Menezes (2005), in terms of the nature of the research, this is an applied study, since it will be carried out with the aim of generating knowledge that will then be applied to the company.

According to Silva and Menezes (2005) in terms of approach, it is classified as quantitative research, as the results will be translated into numbers and some statistical techniques will be used, such as mean, median, mode, standard deviation, among others.

According to Gil (2007), this work is considered to be descriptive research, as it aims to describe the characteristics of a particular population or phenomenon or to establish relationships between variables. It involves the use of standardised data collection techniques: questionnaires and systematic observations.

According to Gil (2007), this work is also classified as a case study, as an in-depth study of the methods used in the company will be carried out, along with a study of the processes for making the product, seeking broad and detailed knowledge.

Below are the steps for carrying out this work:

- Literature review using books on the subject of this work;
- Observe and characterise the sector under study;
- Analyse the working method used;
- Map the activities of the sector and each operator using the flowchart and notes;
- Develop a timing sheet;
- Collect the times;
- Calculate standard time and production capacity;
- Identifying improvements and proposing new working methods;
- Apply a new method;

- Collect new times;
- Calculate new standard time and new production capacity;
- Compare the standard times and quantify the gains.

CHAPTER 4

DEVELOPMENT

This topic will cover the company in general, its management systems and, in particular, the foaming sector, describing the production process analysed.

4.1. Company characterisation

The company under study began with a project in 1999 and was founded on 30 March 2000 in the city of Maringá, in the north-west of the state of Paraná, operating in the magnetic mattress sector. It is currently the market leader in its segment, with branches strategically spread throughout the country (Figure 2), and has its own fleet of trucks to deliver its products to the branches. It has around 600 employees and its factory has an area of approximately 39900 metres^2 and is currently expanding.

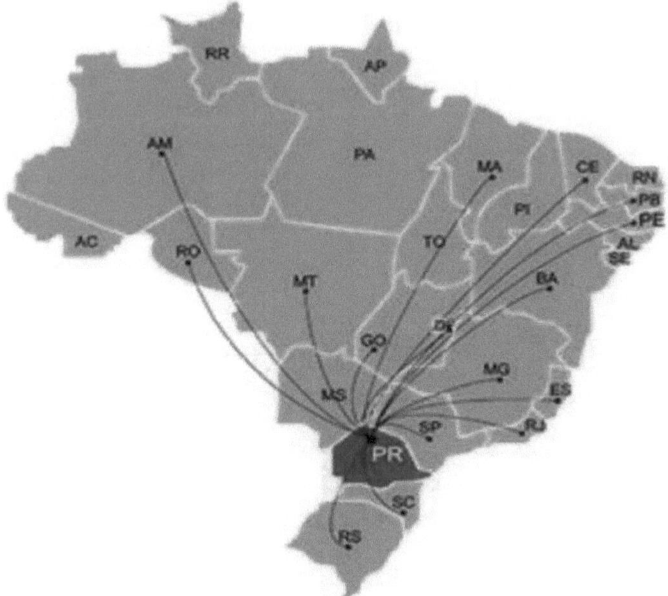

Figure 2: States where the company has branches
Source: Company (2014)

Today the company is a benchmark for quality in the manufacture of its products, exporting to various countries in South America, North America, Europe and Africa, as it invests its resources in order to provide rest, health and well-being for its customers and business opportunities for people who want to become entrepreneurs.

The company has a quality management system with ISO 9001:2008 certification, obtained in November 2004 by Bureau Veritas Certification. This quality system includes procedure manuals, work instructions, parameter tables and performance indicators, as well as an integrated management system. And to complete its quality management system, the company has its products certified by INMETRO.

The company also has social and environmental responsibilities. It has an Environmental Management system with ISO 14001 certification, which it achieved in December 2007 through a lot of effort and investment in infrastructure and training, making it the only company in the magnetic mattress industry to have this certification. The company promotes environmental actions such as: environmental awareness and education through periodic training for employees and their families, holding events on commemorative dates (International Environment Day, Tree Day, Water Day, etc.), practising selective waste collection within the company and encouraging employees to do the same at home. It runs programmes to reduce water and electricity consumption.

Also in terms of environmental management, the company manages its solid waste and 100% of the waste generated is properly disposed of by legal and specialised companies. Figure 3 shows the destination of the waste generated.

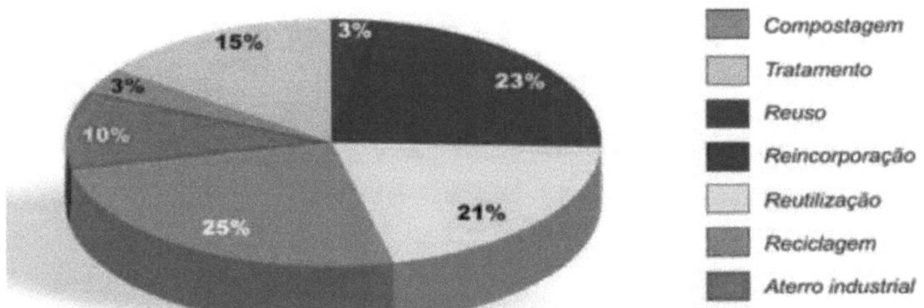

Figure 3: Percentage of waste destinations Source: Company (2014)

In order for the company to be able to segregate its waste, there are coloured selective bins throughout the company, as shown in Figure 4, according to their characteristics, and in each sector there are specific bins for the waste generated by each sector.

Figure 4: Coloured recycling bins Source: Company (2014)

In addition to solid waste management, the company also manages liquid waste, which is duly disposed of in waterproof containment boxes and then sent for treatment by a specialised company duly regulated by the competent authorities. It also manages atmospheric emissions.

To complete the select group of certifications, the company has a Federal Operating Authorisation granted by ANVISA / MINISTRY OF HEALTH, which was granted on 13 September 2010, making its products of increasingly high quality for its customers. This certification was due to the fact that the magnetic mattress offers health benefits to customers.

With all these certifications achieved by the company and with its growth, the company has a hierarchical level with various sectors and departments, as shown in Annex A, in order to

4.2. Characterisation of the Foaming Sector

The foaming sector is responsible for manufacturing polyurethane foam blocks, i.e. the main raw material for mattresses, and is therefore one of the company's main sectors. The company produces five different densities, D20, D28, D33, D40 and D50 (the letter "D" represents that the density produced is real, i.e. after the blocks are manufactured, the density of each foam produced is calculated by dividing the mass by the volume), each density has a specific colour established by the company itself and is described in the instruction manual. Figure 5 illustrates the structure of two magnetic mattress models, showing the colour of each density and where each density is placed in each model, with the D100 Foam Agglomerate layer and the Orthopaedic Box being supplied by third parties.

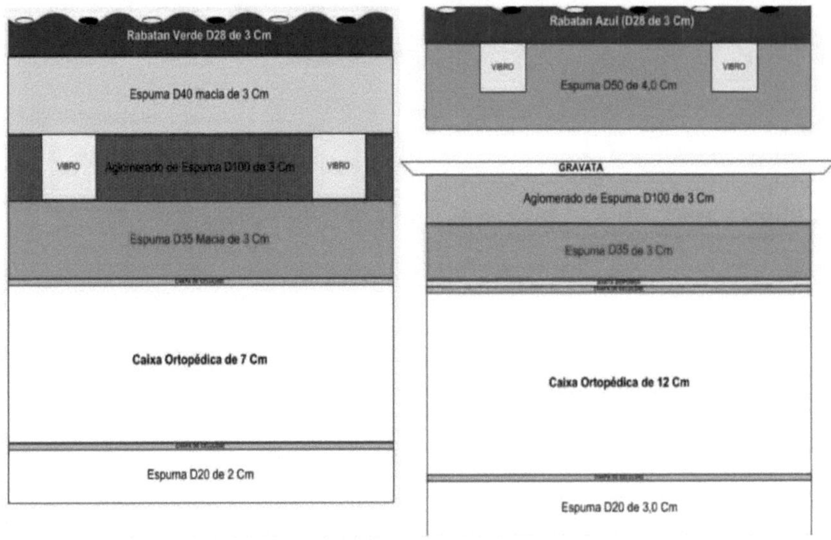

Figure 5: Structures of two magnetic mattress models
Source: Company (2014)

This sector has a production leader, a foamer and four other assistants, as shown in Figure 6. The leader is responsible for receiving the daily programme and passing on the quantity of foam blocks of each density to be produced each day to the employees.

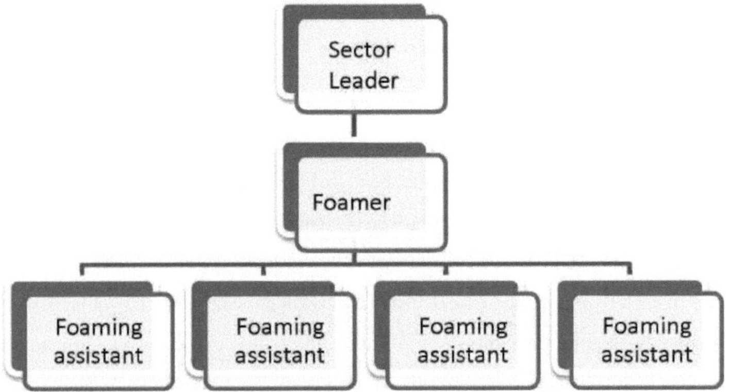

Figure 6: Organisational chart of the foaming sector Source: Author (2014)

The manufacture of the polyurethane foam block, like any production process, has its input, which are the inputs, the process and the output, which is the polyurethane foam block. Figure 7 shows this description.

Figure 7: Manufacturing process for the polyurethane foam block.
Source: Author (2014)

Polyurethane foams are manufactured using exclusive formulas developed by a chemical engineer for the company, where each foam density receives a specific quantity of chemical products. Copolymers (high value-added raw materials) are used to guarantee the superior quality of the product, and inspections are carried out during and after the process. The history of these inspections is kept in order to enable product traceability and ensure that the density produced is real. At the end of manufacturing, issues such as: block relaxation, textures and temperature (ambient temperature at which the foam block was produced) are checked. The polyurethanes used are tested and approved by laboratories for safety and responsibility towards the consumer in order to guarantee the product's 15-year guarantee. To manufacture the foam blocks, the company has its own semi-automatic machine (Figure 8) in which the main products used (TDI, copolymer and polyol) are programmed by the foamer for automatic weighing for each density to be manufactured.

Figure 8: Foam-making machine Source: Company (2014)

For a better understanding of the foaming machine, the parts highlighted in the Figure are characterised in the Table

1. **Machine control:** Where the foamer programmes the weighing of the chemical products (polyol,

copolymer and TDI) for each density;

2. **Beater:** This cylinder is used to pour in the chemicals and manually weighed products (water, amine, silicone and tin) to carry out the 1^{a}, $2^{(a)}$and 3^{a} beats;
3. **Polyol and copolymer container:** This container holds the polyol and copolymer together, as programmed by the foamer to make the foam block;
4. **TDI container:** Where the TDI is stored, as programmed by the foamer for making the foam block;
5. **TDI container:** Here the TDI flows from the container (4) in the right quantity, according to the foamer's programming, to make the block that will be produced and from this container, the TDI is poured into the beater to make the 3^{a} Beat;
6. **Box:** Where the products mixed by the mixer are poured in and the foam block begins to take shape;
7. **Lid:** After the products have been poured into the box, this lid goes on top of the box and an assistant lowers it manually and the gases expelled by the chemical reaction come out through the pipe attached to this lid.

To weigh the other products (water, silicone, amine and tin), the company has a chapel (Figure 9) which has a duct to evacuate the product gases, a scale to weigh the products and a thermometer, as the quantity of these products varies with the temperature when making foam blocks. Next to the chapel is a cupboard for storing dyes, as each foam density has a specific colour.

Figure 9: Chapel for weighing products Source: Author (2014)

4.3 Chronoanalysis

Before starting the chrono-analysis, the sector leader and employees were informed about what the chrono-analysis was and how it would be carried out so that everyone was aware that the time measurement would focus on the processes and not the employees themselves. This prior sensitisation promoted a high level of understanding and collaboration from the whole team.

Even before the chrono-analysis, the sector was analysed (area of the shed, raw material intake, movement of employees) and how the foam blocks were produced, the densities produced, where they were stored, and knowledge of the entire production process and what each employee did, in order to develop the best method for timing the foaming process and to analyse possible improvements in the sector and in the employees' working methods. After analysing the sector, a flowchart was drawn up of the activities involved in the process, as shown in Figure 10.

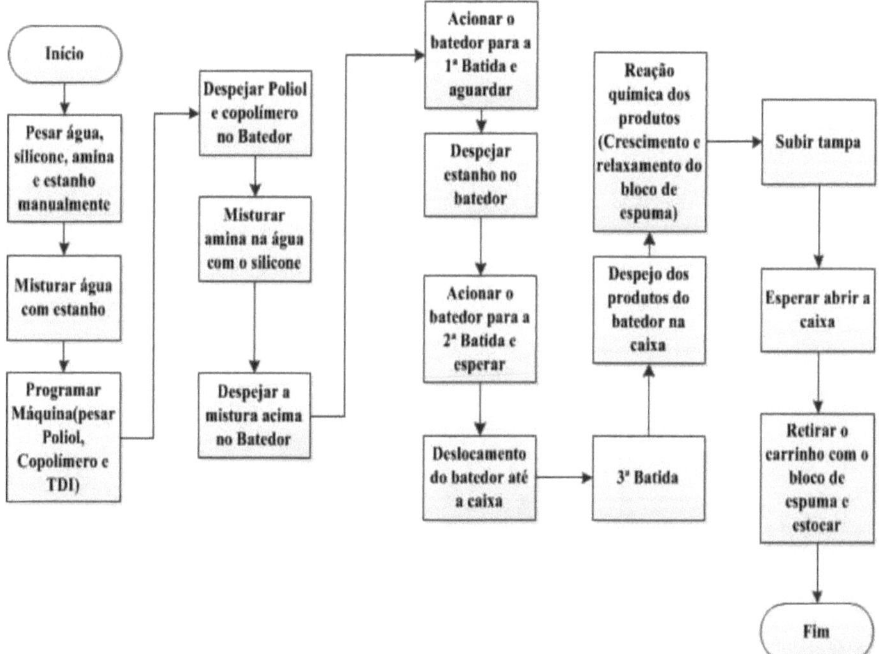

Figure 10: Flowchart of the foaming process Source: Author (2014)

A timing sheet was then developed (Figure 11) which was used to collect the times of the activities carried out in the foam manufacturing process. Its structure contains the following information: the sector where the timing was being carried out, the density of the foam block to be produced, the date on which the time measurement was carried out, the shift in which the foam block production time was collected, the number of employees involved in the process, the elements to be timed, a field for the necessary observations involved in the process and the start and end times of each element, since the continuous reading method was used to collect the time of each element, according to Barnes (2008).

Timing Sheet				
Sector:				
Density:		**Date:**		
Number of Employees		**Shift:**		
		Times		
N	**Elements**	**initial t**	**t final**	**Duration**
1				
2				
3				
4				
5				
6				
7				
8				
9				
10				
11				
12		**Total time**		
OBS:				

Figure 11: Timing sheet Source: Author (2014)

The following materials were used to carry out the chronoanalysis: clipboard, timing sheet, pen, chronometer, however due to the difficulty of being able to find a centesimal chronometer an ordinary chronometer was used, and before the timing was carried out the chronometer was calibrated as required by the ISO 9001 standard, which states that measuring instruments must be calibrated and the sticker with the barcode affixed to the chronometer (Figure 12) is for its traceability, where it will indicate which company carried out the calibration, which was calibrated, which day it returned, when it was acquired, among other information.

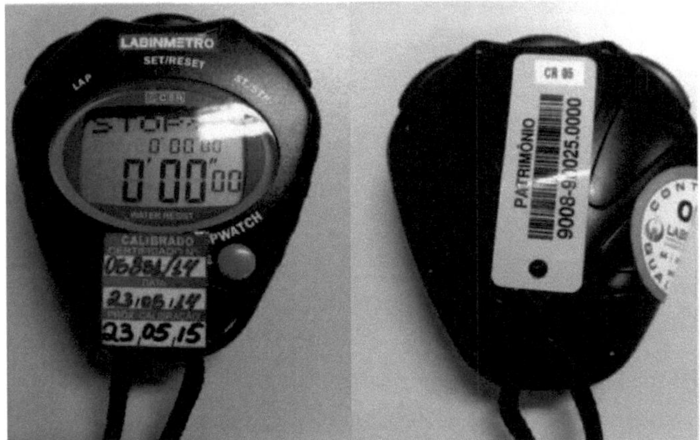

Figure 12: Calibrated stopwatch

Source: Author (2014)

To carry out the timing, the foam manufacturing process was divided into the following time elements, as shown in Table 5.

Elements	Description	Image
Weighing products by hand	Process carried out by the foamer, in which he weighs the water, silicone, amine and tin and mixes the silicone with the water;	
Walking from the chapel to the machine	Moving the foamer from the chapel to the machine.	
Programming the machine	Process carried out by the foamer, where the machine is programmed to weigh the polyol and copolymer for the next block;	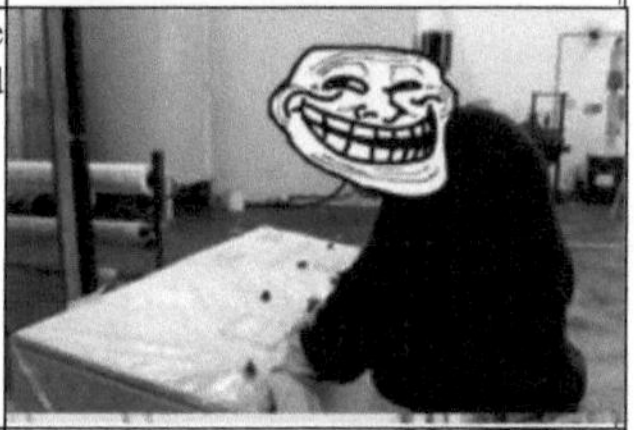
Floor of machine to the chapel	Moving the foamer from the machine to the chapel;	
Mix Amine	Process carried out by the foamer, where amine is mixed with water and silicone;	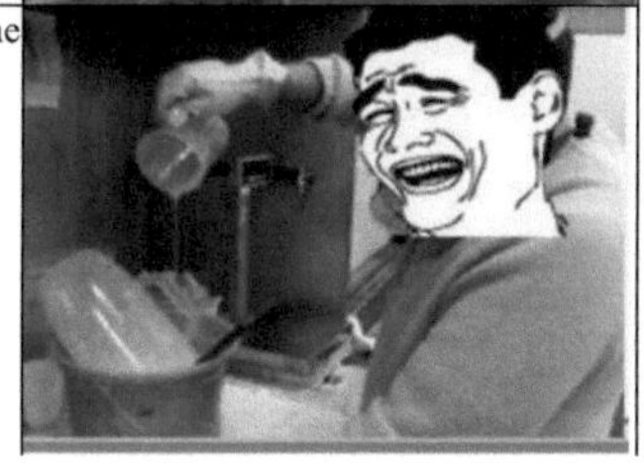

Pour the products weighed manually in the mixer	These are processes that involve moving from the chapel to the machine, carrying the bucket with the mixture of water with silicone and amine, programming the machine for the beats and finally pouring the products into the beater; 
1ª Hit + 2ª Hit	This process consists of the 1ª and 2ª strokes of the beater. When the E beat is taking place, the foamer moves from the machine to the chapel, picks up the tin, returns to the machine and pours it into the beater. E also involves moving the beater to the "box" where the foam block grows; 
3ª Beat + growth + relaxation	This process involves the machine pouring the TDI into the beater, the third beat, pouring the beater products into the "box", the chemical reaction to grow the foam block, relaxing the foam block and opening the lid of the "box" to remove the foam block; 
Remove foam block	Process that involves taking the trolley with the foam block to the place to store the foam block. 

Chart 5: Elements, description and images of the foaming process activities
Source: Author (2014)

For a better understanding of the flow of the procedure, Figure 13 illustrates the layout of the foaming sector.

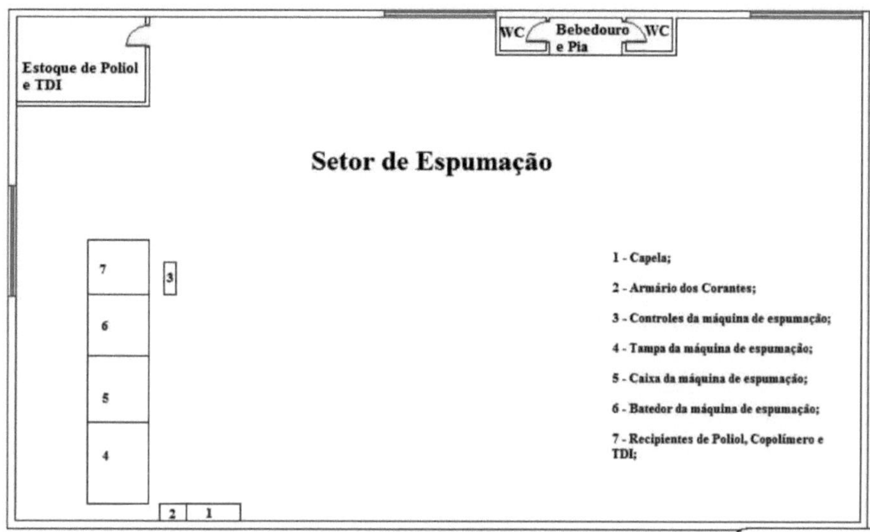

Figure 13: Sketch of the foaming sector Source: Author (2014)

After creating the timing sheet and dividing the process into elements, the timing began. Initially, 7 measurements were taken for each density in order to estimate the sufficient number of measurements to be taken using Equation 1. For this, a 90% probability was adopted for the degree of reliability of the measurement, with a normal distribution coefficient for a given probability Z = 1.65 as shown in Table 1, the acceptable relative error (e) was 10%. As the initial number of measurements was 7, the coefficient as a function of the number of preliminary timings (d_2) is 2.704, as shown in Table 2. Tables 4 to 8 show the times for each density, the average and the amplitude in centesimal time; the conversion was made according to the example in Table 1.

Table 4: D20 Density times, average and amplitude

D20 density								
Times							Average	Amplitude
10,82	12,7	9,0	9,13	10,5	9,8	9,55	**10,21**	**3,7**

Source: Author (2014)

Table 5: Density D28 times, average and amplitude

Density D28								
Times							Average	Amplitude
11,32	11,67	12,42	13,82	10,77	11,42	10,65	**11,64**	**3,17**

Source: Author (2014)

Table 6: D33 Density times, average and amplitude

D33 density								
Times							Average	Amplitude
14,82	13,68	13,47	12,12	13,62	10,97	11,1	**12,83**	**3,85**

Source: Author (2014)

Table 7: D40 Density times, average and amplitude

Density D40								
Times							Average	Amplitude
14,13	14,9	14,4	11,8	11	10,8	12,17	**12,74**	**4,1**

Source: Author (2014)

Table 8: Density D50 times, average and amplitude

Density D50								
Times							Average	Amplitude
17,67	16,25	15,27	14,27	15,65	18	16,38	**16,21**	**3,73**

Source: Author (2014)

After taking the time measurements, the number of cycles to be timed for each density was calculated according to Equation 1. Table 9 shows the calculations for each density and the number of timings required for each density.

Table 9: Calculating the number of timings for each density

Density	Calculation	No. of timings (n)
D20	$n = \left(\frac{1,65 * 3,7}{0,1 * 2,704 * 10,21}\right)^2 = 4,89$	5
D28	$n = \left(\frac{1,65 * 3,17}{0,1 * 2,704 * 11,64}\right)^2 = 2,76$	3
D33	$n = \left(\frac{1,65 * 3,85}{0,1 * 2,704 * 12,83}\right)^2 = 3,35$	4
D40	$n = \left(\frac{1,65 * 4,1}{0,1 * 2,704 * 12,74}\right)^2 = 3,86$	4
D50	$n = \left(\frac{1,65 * 3,73}{0,1 * 2,704 * 16,21}\right)^2 = 1,97$	2

Source: Author (2014)

As the number of timings to be carried out varied from 2 to 5, a standard of 5 timings was established for all densities. After calculating the number of timings, a new timing table was developed (Figure 14).

Timing Sheet							
Sector:							
Density:				**Date:**			
Number of Employees				**Shift:**			
X	**Elements**	**Times**					**Average**
		11	**t2**	**t3**	**14**	**t 5**	
1							
2							
3							
4							
5							
6							
7							
S							
9							
10							
Total Time							
Speed (Rhythm)							
Normal Time							

Standard Time	

Figure 14: Timing sheet
Source: Author (2014)

When timekeeping work is carried out and the employee is aware that they are being evaluated, it is normal for there to be a change in their working pace, either to a fast or slow pace. Assessing the speed at which the employee is working is one of the most difficult and important parts of chrono-analysis. However, as the company had already carried out chrono-analysis in previous years, the employees were already used to measuring the process. And since the employees were told about this study and the importance of carrying out activities at a pace that is considered normal (not fast or slow), Table 3 was used to determine the RR.

Skill = Medium D = +0.00;

Effort = Medium D = +0.00;

Conditions = Average D = 0.00;

Consistency = Average D = +0.00

Total = +0.00 + 0.00 + 0.00 + 0.00 = 0.00 + 1 = 1.00 = 100%

Therefore, in this case the FR to be considered was 100 per cent, so the Timed Time will be equal to the Normal Time.

Once Normal Time has been determined, it must be taken into account that a worker cannot work all day without any breaks, such as for personal needs, fatigue or other reasons. The foaming sector works from 7.30am to 5.30pm, with a 1 hour 12 minute break for lunch (from 11.18am to 12.30pm), so it has a daily working day of 8 hours 48 minutes (528 minutes), but there are 15 minutes for gymnastics and 15 minutes for afternoon coffee every day, as well as preparation of the sector, which results in 2 hours a day. So the Tolerance Factor was determined as follows: FT = (personal needs = 5%; fatigue = 10%; gym and coffee = 6%; preparation time = 22%; cleaning = 5%). Therefore, the FT represents 48 per cent of the time available for work.

This information was used to determine the average, normal time and standard time for each density, as shown in Figures 15 to 19, followed by a comparison between the averages for each density, thus determining an overall average and an overall standard time, as shown in Figure 20, in order to calculate the production capacity of the foaming sector.

Timing Sheet							
Sector: Foaming							
Density: D20		**Date: 03/06/14 to 04/06/14**					
Number of employees: 5		**Shift:** Evening					
N	**Elements**	**Times (in minutes)**					**Average**
		tl	**t2**	**t3**	**t4**	**t 5**	
1	Weighing products by hand	1,75	1,83	2,12	2,00	2,03	1,95

2	Walking from the chapel to the machine	0,30	0,17	0,15	040	0,17	040
3	Machine programming	1,83	135	142	1,05	0,97	144
4	Walking from the machine to the chapel	0,15	0,17	0,17	0,17	0,13	0,16
5	Mixing amma	043	035	0,40	038	042	048
ô	Pour the products into the mixer	0,33	0,38	0,37	038	0,35	0,36
7	Iª Beat - 2ª Beat	1,37	1,40	1,47	1.52	1.50	1.45
8	3ª Beat - growth - relaxation	4,00	4,42	3,78	**442**	7,30	4,74
9	Remove foam block	0,13	0,17	0,13	0,18	0,12	0,15
Total Time		**10,09**	**10,04**	**9,71**	**10,00**	**12,79**	**10,53**
Speed (Rhythm)							100%
Normal Time							10,53
Tolerance Factor							1,48
Standard Time							**15,58**

Figure 15: D20 density times
Source: Author (2014)

Timing Sheet							
Sector: Foaming							
Density: D28				**Date:** 04/06/14			
Number of employees: 5				**Shift:** Evening			
N	Elements	Times (in minutes)					Average
		tl	t2	t3	t 4	t 5	
1	Weighing products by hand	1,78	2,00	1,70	2,33	1J8	1,92
2	Walking from the chapel to the machine	0,12	0,15	0,13	0,15	020	0,15
3	Machine programming	1,37	0,97	1,07	0,92	1,92	125
4	Walking from the machine to the chapel	0,12	0,15	0,17	0,18	0,15	0,15
5	Mixing animation	0,17	0,18	0,23	027	0,12	0,19
6	Pour the products into the mixer	0,40	0,37	0,42	0,40	0,50	0,42
7	Iª Hit -2ª Hit	0,77	0,85	0,78	1,15	1,03	0,92
8	3ª Beat + growth+ relaxation	5,78	6,90	5,53	6,02	6,53	6,15
9	Remove foam block	0,53	0,63	0,63	0,62	0,65	**0,61**
Tempo Tota!		**11,04**	**12,20**	**10,66**	**12,04**	**12,88**	**11,76**
Speed (Rhythm)							100%
Normal Time							11,76
Tolerance Factor							1,48
Standard Time							**17,41**

Figura 16: D28 density times
Source: Author (2014)

Timing Sheet							
Sector: Foaming							
Density: D33				**Date:** 04/06/14 to 05/06/14			
Number of employees: 5				**Shift:** Evening			
N	Elements	Times (in minutes)					Average
		11	t2	t3	t 4	t5	
1	Weighing products by hand	1,67	2,87	2,33	1,48	1,57	1,98
2	Walking from the chapel to the machine	0,18	0,17	0,17	022	0,17	0,18
3	Machine programming	0,77	1,82	0,93	0,90	0,38	0,96
4	Walking from the machine to the chapel	0,17	0,13	0,12	0,17	0,15	0,15
5	Mixing amine	0,13	0,22	025	023	022	021
6	Pour the products into the mixer	0,37	0,50	0,53	0,42	0,37	0,44
7	Iª Beat - 2ª Beat	0,70	3,52	1,67	1,65	0,82	1,67
8	3ª Beat - growth+ relaxation	628	4,02	4,80	4,85	5,00	**4,99**
9	Remove foam block	0,58	0,13	020	0,15	0,13	024
Total Time		**10,85**	**13,38**	**11,00**	**10,07**	**8,81**	**10,82**
Speed (Rhythm)							100%
Normal Time							10,82
Tolerance Factor							1,48
Standard Time							**16,02**

Figura 17: D33 density times

Source: Author (2014)

Timing Sheet							
Sector: Foaming							
Density: D40				**Date:** 05/06/14			
Number of employees: 5				**Shift:** Afternoon			
N	**Elements**	**Times (in minutes)**					**Average**
		11	**t2**	**t 3**	**t 4**	**15**	
1	Weighing products by hand	1=47	1,63	1,43	1,72	1,52	1=55
2	Walk from the chapel to the machine	**020**	0,17	017	0,17	**0,15**	0,17
3	Machine programming	1,88	1= 80	1,00	1,57	1,63	1=58
4	Walking from the machine to the chapel	0,17	0,13	017	017	0,12	015
5	Mixing animation	U7	025	023	028	025	0,46
6	Pour the products into the mixer	0,47	057	042	0,47	0,42	0,47
7	Iª Beat - 2ª Beat	1,63	1,50	1,62	1,70	1=55	1,60
8	3ª Beat - growth - relaxation	4,85	5,43	5,80	5,33	6,88	5,66
9	Refurbishing the foam block	023	0,15	017	0,17	0,18	0,18
Total Time		**12 J 7**	**11,63**	**11,01**	**11,58**	**12,70**	**11,82**
Speed (Rhythm)							100%
Normal weather							11,82
Tolerance Factor							1,48
Standard Time							**17,49**

Figura 18: D40 density times
Source: Author (2014)

Timing Sheet							
Sector: Foaming							
Density: 050				**Date:** 03/06/14 to 05/06/14			
Number of employees: 5				**Shift:** Evening			
N	**Elements**	**Times (in minutes)**					**Average**
		11	**t2**	**t3**	**t4**	**t 5**	
1	Weighing products by hand	2,67	2,35	225	227	2,15	2,34
2	Walking from the chapel to the machine	0,13	0,15	**0,17**	0,12	0,18	0,15
3	Machine programming	1,00	0,97	1=13	1,78	2,00	1,38
4	Walk from the machine to the chapel	0,10	0,13	**0,17**	0,13	0,13	0,13
5	Mixing amine	022	0,12	0,33	0,13	027	**021**
6	Pour the products into the mixer	0,95	0,98	0,75	0,67	0,60	0,79
7	Iª Beat - 2ª Beat	023	0,50	0,38	0,50	1,35	0,59
8	3ª Beat - growth - relaxation	8,78	8,43	8,67	9,65	9,42	8,99
9	Remove foam block	2,17	1,63	0,42	0,40	1,90	1,30
Total Time		**1625**	**1526**	**1427**	**15,65**	**18,00**	**15.89**
Speed (Rhythm)							100%
Normal weather							15,89
Tolerance Factor							1,48
Standard Time							**23,51**

Figura 19: D50 density times
Source: Author (2014)

Comparison of density averages and calculation of overall Standard Time							
N	**Elements**	**Densities**					**Average**
		D20	**D28**	**D33**	**D+0**	**D50**	
I	Weighing products by hand	1,95	**1,92**	1,98	1,55	2,34	1,95
2	Walking from the chapel to the machine	0,20	0,15	0,18	0,17	0,15	0,17
3	Machine programming	1,24	1,25	0,96	1,58	1,38	1,28
4	Walk from the machine to the chapel	0,16	0,15	0,15	0,15	0,13	0,15
5	Mixing amine	0,28	0,19	0,21	0,46	0,21	0,27
6	Pour the products into the mixer	0,36	0,42	0,44	0,47	0,79	0,50
7	1* Shake + 2* Shake	1,45	0,92	1,67	1,60	0,59	U5

B	3ª Beat + growth + relaxation	4,74	6,15	4,99	5,66	8,99	6,11
9	Remove foam block	0,15	0,61	024	0,18	1,30	0,50
	Total Time	**10,53**	**11,76**	**10,82**	**11,82**	**15,89**	**12,16**
	Speed (Rhythm)						100%
	Normal Time						12,16
	Tolerance Factor						1,48
	Standard Time						**18,00**

Figura 20: Comparison between the times of each density
Source: Author (2014)

After calculating the overall Standard Time, the sector's production capacity was calculated using Equation 6.

$$C = \frac{HT}{TP} \tag{6}$$

Where:

C = Production capacity;

HT = Hours worked (528 minutes);

TP = Standard Time (18 minutes).

$$C = \frac{528}{18} = 29{,}33 \text{ unidades}$$

Therefore, the sector's production capacity is 29 foam blocks per day, but it can only produce 25 foam blocks in its working day (from 7.30am to 5.30pm), so it has to work two hours of overtime a day to produce around 35 foam blocks.

4.4. Improvements and application of the new method

In addition to determining standard time and production capacity, chronoanalysis allows us to observe and analyse the sector as a whole, analysing each operator, the flow of the production process, ergonomics, layout, among other factors that improve the entire production system.

After carrying out the chrono-analysis, some ideas for improvement were proposed by analysing the timed activities, the observations made in relation to the movement of employees to carry out the process and the times obtained to reduce daily overtime and increase the productive capacity of this sector:

- Weighing the products (water, amine, silicone and tin) in parallel with the block growth and relaxation stage, as the quantities of these products rarely varied for the manufacture of the same foam block, for this idea it was suggested to train one or more assistants to weigh these products;
- Start the working day earlier, as electricity costs more after 6pm, or send a team in at 6am to prepare the sector;
- Place the chapel in front of the beater, about 2.5 metres away, to reduce movement between the chapel

and the beater, as they are approximately 8 metres apart;

- Place the cupboard, which holds the colourants, next to the chapel and place it behind the whisk to reduce the distance.

These ideas were written down in a report, shown and discussed with the sector leader, the production coordinators and the production manager to see if it would be possible to make these improvements. Of the ideas proposed, only the idea of placing the chapel in front of the mixer was not accepted, because the company is expanding its physical structure, building new sheds and the foaming sector will be moved to the shed next door, so it would not be feasible to make this change, which would take a lot of time to change again later on.

In order to put the ideas for improvements into practice, the leader met with the sector's employees and talked to them, explaining what was going to be done and why it was being done, and soon the employees accepted the new proposals and collaborated with the changes. The leader was in charge of selecting two assistants to be trained to weigh the products (water, amine, silicone and tin) and together with the production coordinator they decided to select three employees to come in at 6am and leave at 4pm to prepare the foaming sector and another cupboard was placed behind the beater to store the colourants, as the previous cupboard wouldn't fit, Figure 21 shows the new cupboard and Figure 22 illustrates the new **layout.**

Figure 21: New cupboard for storing colourants Source: Author (2014)

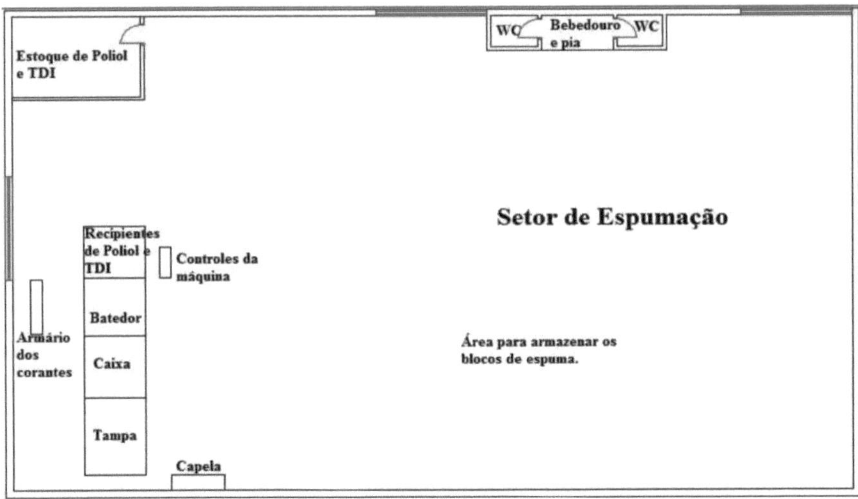

Figure 22: Sketch of the new layout of the foaming sector Source: Author (2014)

The implementation of these improvements led to a new working method, so in order to carry out the chronoanalysis again, a period of one month was given until the employees were used to it and trained to be able to measure the process. After the month of adaptation, the sector was analysed and the foam manufacturing process was divided into the following elements, as shown in Chart 6.

Elements	Description	Image
Put the trolley in the box and wait for it to close	A process where the employee places the trolley in the box, presses a button on the machine's control panel to close it and waits for it to close completely;	
Walk from the chapel to the machine and throw products into the mixer	Moving the employee who weighed the products (water, amine, silicone and tin) from the chapel to the mixer with the bucket and pouring the products into the mixer;	

Iª Hit	The process of mixing (Iª tapping) the manually weighed products with the products in the mixer, the part of picking up tin takes place in parallel to the Iª tapping;	
2ª Beat	Process where the employee pours the tin into the beater and the 2ª Beat takes place;	
3ª Beat + growth + relaxation	The process in which the 3ª beat takes place, the products are poured into the box, and the block grows and relaxes;	
Waiting for the till to open	Process where the lid and box are opened to remove the block;	
Remove trolley	Process where the employee removes the trolley and stores the block.	

Chart 6: Elements, description and images after chronoanalysis Source: Author (2014)

With the new division of the elements, the timing of the elements began. Five times were set for the timing of each density, according to the previous timing. The same timing sheet used before the new working method was used to carry out the timing.

In this timing, the same FR (operator speed) was used, which was 100%, so the Normal Time was the same as the Timed Time. To determine the Tolerance Factor, only the sector's preparation time was disregarded, since a team of 3 employees were entering at 6am to do all the preparation, so when the foaming sector employees arrived at 7.30am, the sector was already ready to start manufacturing the foam blocks. As a result, the FT was 26 per cent of the total time. Figures 23 to 27 show the times for each density.

A comparison was made between the averages of each density, thus determining an overall average and an overall standard time, as illustrated in Figure 28, in order to calculate the production capacity of the foaming sector.

Timing Sheet							
Sector: Foaming							
Density: D20				Date: 15/07/14			
Number of employees: 5				Shift: Afternoon			
x	Elements	Times (in minutes)					Average
		tl	t2	t3	t4	t 5	
1	Place trolley and close till	127	0,52	037	0,53	0,55	0,65
2	Walk from the chapel to the machine and throw products into the mixer	035	0,32	025	0,32	033	0,31
3	Iª Hit	038	0,35	0,45	0,35	035	0,38
i	2ª Beat	0,90	0,58	0,53	0,50	0,53	0,61
	3ª Beat+ growth~ relaxation	3,17	323	3,55	3,90	3,52	3,47
6	Waiting for the till to open	0,53	035	0,45	0,47	0,45	0,45
7	Remove trolley	0,60	020	0,18	0,18	020	027
Total Time		720	**5.55**	5,78	625	5,93	6,14
Speed (Rhythm)							100%
Normal Time							6,14
Tolerance Factor							126
Standard Time							7,74

Figura 23: D20 density times after improvements
Source: Author (2014)

Timing Sheet							
Sector: Foaming							
Density: D28				Date: 15/07/14			
Number of employees: 5				Shift: Evening			
x	Elements	Times (in minutes)					Average
		tl	t2	t3	t4	t 5	
1	Place trolley and close till	0,60	0,62	0,50	0,62	0,48	0,56
2	Walk from the chapel to the machine and throw products into the mixer	0,50	0,43	0,30	028	035	0,37
3	Iª Hit	0,43	0,47	0,50	0,55	0,50	0,49
4	2ª Beat	1,05	1,07	1,05	0,67	0,68	0,90
5	3ª Beat+ growth+ relaxation	3,33	3,10	3,18	3,88	4,15	3,53
6	Waiting for the till to open	0,60	0,72	0,63	0,32	0,32	0,52
7	Remove trolley	0,50	0,30	0,50	022	022	0,35
Total Time		7,01	6,71	6,66	6,54	6,70	6,72
Speed (Rhythm)							100%
Normal Time							6,72
Tolerance Factor							126
Standard Time							8,47

Figura 24: D28 density times after improvements
Source: Author (2014)

Timing Sheet							
Sector: Foaming							
Density: D33				Date: 16/07/14			
Number of employees: 5				Shift: Afternoon			
N	Elements	Times (in minutes)					Average
		tl	t2	t3	t4	t5	
1	Place trolley and close till	0.53	0.55	0,55	0,53	0,53	0,54
2	Walk from the chapel to the machine and throw products into	0,28	028	028	027	028	0,28

	the mixer						
3	Iª Hit	0,72	0.77	0,72	0,70	0,70	0,72
4	2ª Beat	1,33	122	120	123	123	124
5	3ª Beat+ growth - relaxation	4.83	4,50	4,42	4.67	4.33	4,55
6	Waiting for the till to open	0,50	0,43	0,58	0,50	0,58	0,52
7	Remove trolley	0,33	0,52	0,58	0,47	0,60	0,50
	Total Time	8,52	827	8,33	837	82=	8,35
	Speed (Rhythm)						100%
	Normal Time						8,35
	Tolerance Factor						126
	Standard Time						10,52

Figura 25: D33 density times after improvements
Source: Author (2014)

Timing Sheet							
Sector: Foaming							
Density: D40			Date: 17/07/14				
Number of employees: 5			Shift: Evening				
N	Elements	Times (in minutes)					Average
		tl	Í2	t3	t4	15	
1	Place trolley and close till	0,58	0,62	0,70	0,63	0,53	0,61
2	Walk from the chapel to the machine and throw products into the mixer	0,52	0,45	127	0,50	2,33	1.01
3	Iª Hit	0,73	0,65	0,70	0,63	0,63	0,67
4	2ª Beat	120	122	1.10	123	1=18	1.19
5	3ª Beat+ growth+ relaxation	5,03	5J2	5,47	5,07	4,92	524
6	Waiting for the till to open	0,62	0,65	0,47	0,54	0,52	0,56
7	Remove trolley	0,50	0,40	0,70	0,40	0,57	0.51
	Total Time	9,18	9,71	10,41	9,00	10,68	9,80
	Speed (Rhythm)						100%
	Normal Time						9,80
	Tolerance Factor						126
	Standard Time						12,34

Figura 26: D40 density times after improvements
Source: Author (2014)

Timing Sheet							
Sector: Foaming							
Density: D50			Date: 17/07/14				
Number of employees: 5			Shift: Afternoon				
N	Elements	Times (in minutes)					Average
		tl	t2	t3	t4	t 5	
1	Place trolley and close till	0,57	0,47	0,80	0,60	0,53	0,59
2	Walk from the chapel to the machine and throw products into the mixer.	0,55	037	0,55	0,50	0,57	0,51
3	Iª Hit	0,72	0,67	0,67	0,52	0,60	0,64
4	2ª Beat	128	1,17	123	123	123	123
5	3ª Beat + growth - relaxation	7,53	7,72	8,90	8,33	8,83	826
6	Waiting for the till to open	0,60	0,55	0,80	0,78	0,60	0,67
7	Remove trolley	0,30	0,38	0,37	0,62	0,50	0,43
	Total Time	11,55	11,33	13,32	12,58	12,86	1233
	Speed (Rhythm)						100%
	Normal Time						12,33
	Tolerance Factor						126
	Standard Time						15,53

Figura 27: D50 density times after improvements
Source: Author (2014)

Comparison of density averages and calculation of overall Standard Time							
x	Elements	Densities					Average
		D20	D28	D33	D 40	D50	

1	Weighing products by hand	0,65	0,56	0,54	0,61	0,59	0,59
2	Walk from the chapel to the machine	0,31	0,37	0,28	1,01	0,51	0,50
3	Machine programming	0,38	0,49	0,72	0,67	0,64	0,58
4	Walk from the machine to the chapel	0,61	0,90	124	1,19	1,23	1,03
5	Mixing amine	3,47	3,53	4,55	524	826	5,01
6	Pour the products into the mixer	0,45	0,52	0,52	0,56	0,67	0,54
7	Iª Beat + 2ª Beat	027	0,35	0,50	0,51	0,43	0,41
Total Time		**6,14**	**6,72**	**8,35**	**9,80**	**1223**	**8,67**
Speed (Rhythm)							100%
Normal Time							8,67
Tolerance Factor							126
Standard Time							**104)2**

Figura 28: Comparison of the times for each density in the new method
Source: Author (2014)

After calculating the overall Standard Time, the sector's production capacity was calculated using Equation 6.

$$C = \frac{528}{10{,}92} = 48{,}35 \text{ unidades}$$

Therefore, the production capacity of the sector has increased to around 48 foam blocks per day, but the sector produces around 40 to 45 foam blocks during its working day (from 7.30am to 5.30pm) depending on daily demand. The sector does not produce the 48 blocks per day calculated because there is no need to exceed the production of 45 blocks per day and because of a lack of space.

In order to quantify the gains that chrono-analysis has made, a table was drawn up to compare the times for each density before and after the new working method, Table 10 shows this comparison.

Table 10: Comparison of times before and after chronoanalysis and reduction of standard time

Density	Average Time (Before)	Average Time (After)	Standard Time (Before)	Standard Time (After)	Reduction in Standard Time (%)
D20	10,53	6,14	15,58	7,74	50,32
D28	11,76	6,72	17,41	8,47	51,35
D33	10,82	8,35	16,02	10,52	34,38
D40	11,82	9,8	17,49	12,34	29,45
D50	15,89	12,33	23,51	15,53	33,94

Source: Author (2014)

By reducing the time it takes to make each density of foam block, there has been a very significant increase in production, from 25 blocks a day to 45, i.e. an 80 per cent increase in production with simple improvements in the sector and by changing the working method.

CHAPTER 5

FINAL CONSIDERATIONS

The literature review on the study of times and methods, productive capacity and ergonomics and work methods made it possible to carry out this work in the foaming sector of a magnetic mattress company. Even though these subjects did not begin in this production environment, the efficiency and flexibility of these themes can be verified, as the study of times and movements began in processes other than this work.

With the theoretical background, it was possible to divide the polyurethane foam block manufacturing process into time elements, making improvements to the working method noticeable, identifying possible activities that could be carried out in parallel to others, as well as analysing the process flow.

After the analyses and time measurements had been carried out, improvements to the working method were proposed in a report for the sector leader, production coordinator, PPCP analyst and production manager to analyse and discuss which of the proposals would be viable or not, and whether in addition to these suggested proposals there could be any other improvements for the sector. Of the proposals suggested, only one was not accepted, and all the others were implemented and accepted without resistance by the sector's employees.

With all this time measurement and analysis, it was possible to reduce the manufacturing time for each foam block and, consequently, increase daily production by 80 per cent (from 25 to 45 blocks per day).

The time and methods study proved to be efficient, achieving the company's objectives, reducing overtime for employees in the sector and increasing the daily production of foam blocks.

Thus, the general and specific objectives of this work were achieved, contributing to a greater understanding of the concepts of the study of time and methods and its importance, with which the company was satisfied and the work of chrono-analysis was requested in all the other sectors of the company.

For future work, we suggest a more detailed study of the preparation of the sector, so that there is no need for staff to enter the working day earlier.

REFERENCES

BARNES, Ralph M.. **Motion and Time Studies:** Design and Measurement of Work. 6. ed. São Paulo: Blucher, 2008. 635 p.

BARRETO, Antônio Amaro Menezes. **Quality and Productivity in the Clothing Industry:** A Question of Survival. 1.ed. Londrina: Midiograf, 1997. 176p.

CAMAROTTO, João Alberto. **Labour Engineering:** Methods, Times, Work Design. Federal University of São Carlos, São Carlos, 2005.

CHIAVENATO, Idalberto. **Introduction to general management theory**. 6 ed. Rio de Janeiro: Campus, 2000.

GIL, Antonio Carlos. **How to prepare research projects.** 4. ed. São Paulo: Atlas, 2007. 175 p.

GOMES, Maria de Lourdes Barreto; MÁSCULO, Francisco Soares. **Work Organisation**. In: MÁSCULO, F. S.; VIDAL, M. C. (Orgs.). **Ergonomics:** Adequate and Efficient Work. Rio de Janeiro: Elsevier/ Abepro, 2011, 648p.

IIDA, Itiro. **Ergonomics:** Design and Production. 2.ed. São Paulo: Blucher, 2005. 614p.

MARTINS, Petrônio G.; LAUGENI, Fernando Piero. **Production Management.** 2. ed. São Paulo: Saraiva, 2005. 562 p.

MÁSCULO, Francisco Soares. **Physical Ergonomics Tools**. In: MÁSCULO, F. S.; VIDAL, M. C. (Orgs.). **Ergonomics:** Adequate and Efficient Work. Rio de Janeiro: Elsevier/ Abepro, 2011, 648p.

PEINADO, Jurandir; GRAEML, Alexandre Reis. **Production Management:** Industrial and Service Operations. 1.ed. Curitiba: Unicenp, 2007. 750p.

SILVA, Edna Lúcia da; MENEZES, Estera Muszkat. **Research Methodology and Dissertation Writing.** Florianópolis, 2005. 139 p. Available at: < https://projetos.inf.ufsc.br/arquivos/Metodologia_de_pesquisa_e_elaboracao_de_teses_e_dissertacoes_4ed.pdf> Accessed on: 26 Apr. 2014.

SLACK, Nigel; CHAMBERS, Stuart; JHONSTON, Robert. **Production Management.** 3. ed. São Paulo: Atlas, 2009. 703 p.

TAYLOR, Frederick W. **Principles of Scientific Management.** 8. ed. São Paulo: Atlas, 1990. 110 p.

TOLEDO Jr., Itys Fides Bueno. **Times and Methods.** 7. ed. Mogi das Cruzes: Raphael A. Godoy, 2004, 181p.

VIDAL, Mario Cesar. **Ergonomic Analysis of Work**. In: MÁSCULO, F. S.; VIDAL, M. C. (Orgs.). **Ergonomics:** Adequate and Efficient Work. Rio de Janeiro: Elsevier/ Abepro, 2011, 648p.

ANNEX A - COMPANY ORGANISATION CHART

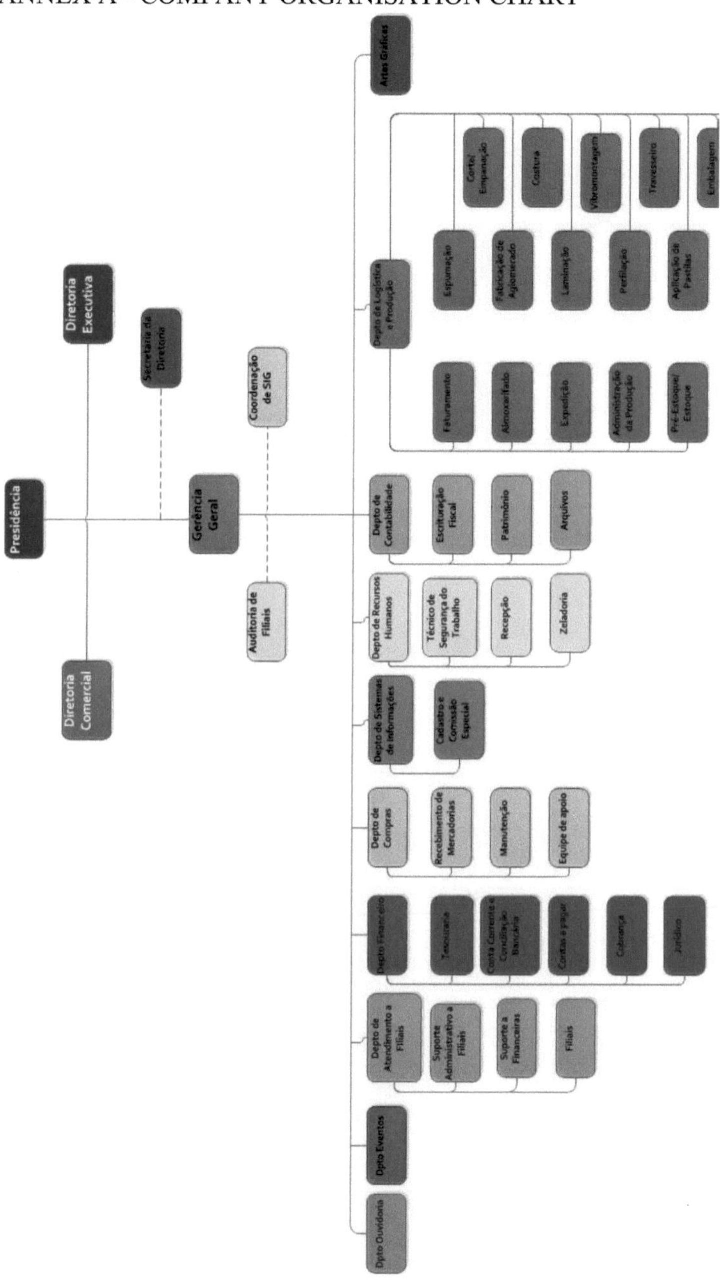

Printed by Books on Demand GmbH, Norderstedt / Germany